EXPERIMENT LOG

Name _____ Address _____

Tel _____ Course _____ Section _____

Instructor _____ Partner(s) _____

Pg #	EXPERIMENT	DATE

Pg #	EXPERIMENT	DATE

EXP. NUMBER	EXPERIMENT/SUBJECT		DATE	
				001
NAME		LAB PARTNER	LOCKER/DESK NO.	COURSE & SECTION NO.

SIGNATURE		DATE	WITNESS/TA		DATE

EXP. NUMBER	EXPERIMENT/SUBJECT		DATE	
NAME		LAB PARTNER	LOCKER/DESK NO.	COURSE & SECTION NO.

001

SIGNATURE		DATE	WITNESS/TA		DATE

001

EXP. NUMBER	EXPERIMENT/SUBJECT		DATE		002
NAME		LAB PARTNER	LOCKER/DESK NO.	COURSE & SECTION NO.	

SIGNATURE		DATE	WITNESS/TA	DATE

EXP. NUMBER	EXPERIMENT/SUBJECT		DATE	
NAME		LAB PARTNER	LOCKER/DESK NO.	COURSE & SECTION NO.

003

SIGNATURE		DATE	WITNESS/TA		DATE

COPY

ACO

EXP. NUMBER	EXPERIMENT/SUBJECT		DATE	004
NAME		LAB PARTNER	LOCKER/DESK NO.	COURSE & SECTION NO.

SIGNATURE	DATE	WITNESS/TA	DATE

EXP. NUMBER	EXPERIMENT/SUBJECT		DATE	005
NAME		LAB PARTNER	LOCKER/DESK NO.	COURSE & SECTION NO.

SIGNATURE		DATE	WITNESS/TA	DATE

EXP. NUMBER	EXPERIMENT/SUBJECT		DATE	
NAME		LAB PARTNER	LOCKER/DESK NO.	COURSE & SECTION NO.

006

EXP. NUMBER	EXPERIMENT/SUBJECT		DATE	
NAME		LAB PARTNER	LOCKER/DESK NO.	COURSE & SECTION NO.

SIGNATURE		DATE	WITNESS/TA		DATE

EXP. NUMBER	EXPERIMENT/SUBJECT		DATE	
NAME		LAB PARTNER	LOCKER/DESK NO.	COURSE & SECTION NO.

SIGNATURE		DATE	WITNESS/TA	DATE

EXP. NUMBER	EXPERIMENT/SUBJECT		DATE		
NAME		LAB PARTNER	LOCKER/DESK NO.	COURSE & SECTION NO.	

007

SIGNATURE		DATE	WITNESS/TA		DATE

EXP. NUMBER	EXPERIMENT/SUBJECT		DATE	
NAME		LAB PARTNER	LOCKER/DESK NO.	COURSE & SECTION NO.

SIGNATURE		DATE	WITNESS/TA		DATE

EXP. NUMBER	EXPERIMENT/SUBJECT		DATE	
				008
NAME		LAB PARTNER	LOCKER/DESK NO.	COURSE & SECTION NO.

SIGNATURE		DATE	WITNESS/TA		DATE

EXP. NUMBER	EXPERIMENT/SUBJECT		DATE	009
NAME		LAB PARTNER	LOCKER/DESK NO.	COURSE & SECTION NO.

SIGNATURE		DATE	WITNESS/TA		DATE

EXP. NUMBER	EXPERIMENT/SUBJECT		DATE		010
NAME		LAB PARTNER	LOCKER/DESK NO.	COURSE & SECTION NO.	

SIGNATURE	DATE	WITNESS/TA	DATE

EXP. NUMBER	EXPERIMENT/SUBJECT		DATE	
NAME		LAB PARTNER	LOCKER/DESK NO.	COURSE & SECTION NO.

SIGNATURE		DATE	WITNESS/TA		DATE

EXP. NUMBER	EXPERIMENT/SUBJECT		DATE		011
NAME		LAB PARTNER	LOCKER/DESK NO.	COURSE & SECTION NO.	

SIGNATURE		DATE	WITNESS/TA		DATE

EXP. NUMBER	EXPERIMENT/SUBJECT		DATE		012
NAME		LAB PARTNER	LOCKER/DESK NO.	COURSE & SECTION NO.	

SIGNATURE		DATE	WITNESS/TA		DATE

EXP. NUMBER	EXPERIMENT/SUBJECT		DATE		012
NAME		LAB PARTNER	LOCKER/DESK NO.	COURSE & SECTION NO.	

SIGNATURE		DATE	WITNESS/TA		DATE
					012

EXP. NUMBER	EXPERIMENT/SUBJECT		DATE	

013

NAME	LAB PARTNER	LOCKER/DESK NO.	COURSE & SECTION NO.

SIGNATURE	DATE	WITNESS/TA	DATE

EXP. NUMBER	EXPERIMENT/SUBJECT		DATE		013
NAME		LAB PARTNER	LOCKER/DESK NO.	COURSE & SECTION NO.	

COPY

SIGNATURE		DATE	WITNESS/TA		DATE

EXP. NUMBER	EXPERIMENT/SUBJECT		DATE	014
NAME		LAB PARTNER	LOCKER/DESK NO.	COURSE & SECTION NO.

SIGNATURE	DATE	WITNESS/TA	DATE

EXP. NUMBER	EXPERIMENT/SUBJECT		DATE	
				014
NAME		LAB PARTNER	LOCKER/DESK NO.	COURSE & SECTION NO.

SIGNATURE		DATE	WITNESS/TA		DATE

EXP. NUMBER	EXPERIMENT/SUBJECT		DATE	015
NAME		LAB PARTNER	LOCKER/DESK NO.	COURSE & SECTION NO.

SIGNATURE		DATE	WITNESS/TA		DATE

EXP. NUMBER	EXPERIMENT/SUBJECT		DATE	
NAME		LAB PARTNER	LOCKER/DESK NO.	COURSE & SECTION NO.

015

SIGNATURE		DATE	WITNESS/TA		DATE

EXP. NUMBER	EXPERIMENT/SUBJECT		DATE	
NAME		LAB PARTNER	LOCKER/DESK NO.	COURSE & SECTION NO.

SIGNATURE		DATE	WITNESS/TA	DATE

EXP. NUMBER	EXPERIMENT/SUBJECT		DATE		016
NAME		LAB PARTNER	LOCKER/DESK NO.	COURSE & SECTION NO.	

SIGNATURE		DATE	WITNESS/TA		DATE

EXP. NUMBER	EXPERIMENT/SUBJECT		DATE	
NAME		LAB PARTNER	LOCKER/DESK NO.	COURSE & SECTION NO.

SIGNATURE		DATE	WITNESS/TA		DATE

EXP. NUMBER	EXPERIMENT/SUBJECT		DATE	
NAME		LAB PARTNER	LOCKER/DESK NO.	COURSE & SECTION NO.

SIGNATURE		DATE	WITNESS/TA		DATE

EXP. NUMBER	EXPERIMENT/SUBJECT		DATE		018
NAME		LAB PARTNER	LOCKER/DESK NO.	COURSE & SECTION NO.	

SIGNATURE		DATE	WITNESS/TA		DATE
					018

EXP. NUMBER	EXPERIMENT/SUBJECT		DATE	
NAME		LAB PARTNER	LOCKER/DESK NO.	COURSE & SECTION NO.

SIGNATURE		DATE	WITNESS/TA		DATE

EXP. NUMBER	EXPERIMENT/SUBJECT		DATE	
NAME		LAB PARTNER	LOCKER/DESK NO.	COURSE & SECTION NO.

SIGNATURE		DATE	WITNESS/TA		DATE

EXP. NUMBER	EXPERIMENT/SUBJECT		DATE	
NAME		LAB PARTNER	LOCKER/DESK NO.	COURSE & SECTION NO.

SIGNATURE		DATE	WITNESS/TA	DATE

EXP. NUMBER	EXPERIMENT/SUBJECT		DATE	
NAME		LAB PARTNER	LOCKER/DESK NO.	COURSE & SECTION NO.

SIGNATURE		DATE	WITNESS/TA		DATE

EXP. NUMBER	EXPERIMENT/SUBJECT		DATE	
NAME		LAB PARTNER	LOCKER/DESK NO.	COURSE & SECTION NO.

021

SIGNATURE		DATE	WITNESS/TA		DATE

EXP. NUMBER	EXPERIMENT/SUBJECT		DATE	
NAME		LAB PARTNER	LOCKER/DESK NO.	COURSE & SECTION NO.

022

SIGNATURE		DATE	WITNESS/TA		DATE

270

EXP. NUMBER	EXPERIMENT/SUBJECT		DATE		022
NAME		LAB PARTNER	LOCKER/DESK NO.	COURSE & SECTION NO.	

SIGNATURE		DATE	WITNESS/TA		DATE

EXP. NUMBER	EXPERIMENT/SUBJECT		DATE	
NAME		LAB PARTNER	LOCKER/DESK NO.	COURSE & SECTION NO.

023

SIGNATURE		DATE	WITNESS/TA		DATE

023

EXP. NUMBER	EXPERIMENT/SUBJECT		DATE	
NAME		LAB PARTNER	LOCKER/DESK NO.	COURSE & SECTION NO.

024

SIGNATURE		DATE	WITNESS/TA		DATE

024

EXP. NUMBER	EXPERIMENT/SUBJECT		DATE	
NAME		LAB PARTNER	LOCKER/DESK NO.	COURSE & SECTION NO.

024

SIGNATURE	DATE	WITNESS/TA	DATE

EXP. NUMBER	EXPERIMENT/SUBJECT		DATE	
NAME		LAB PARTNER	LOCKER/DESK NO.	COURSE & SECTION NO.

SIGNATURE		DATE	WITNESS/TA		DATE

EXP. NUMBER	EXPERIMENT/SUBJECT		DATE		025
NAME		LAB PARTNER	LOCKER/DESK NO.	COURSE & SECTION NO.	

SIGNATURE		DATE	WITNESS/TA		DATE
					025

EXP. NUMBER	EXPERIMENT/SUBJECT		DATE		026
NAME		LAB PARTNER	LOCKER/DESK NO.	COURSE & SECTION NO.	

SIGNATURE	DATE	WITNESS/TA	DATE

EXP. NUMBER	EXPERIMENT/SUBJECT		DATE		026
NAME		LAB PARTNER	LOCKER/DESK NO.	COURSE & SECTION NO.	

SIGNATURE	DATE	WITNESS/TA	DATE

EXP. NUMBER	EXPERIMENT/SUBJECT		DATE	
NAME		LAB PARTNER	LOCKER/DESK NO.	COURSE & SECTION NO.

027

SIGNATURE		DATE	WITNESS/TA		DATE

EXP. NUMBER	EXPERIMENT/SUBJECT		DATE	
NAME		LAB PARTNER	LOCKER/DESK NO.	COURSE & SECTION NO.

SIGNATURE		DATE	WITNESS/TA	DATE

EXP. NUMBER | EXPERIMENT/SUBJECT | DATE

028

NAME | LAB PARTNER | LOCKER/DESK NO. | COURSE & SECTION NO.

SIGNATURE | DATE | WITNESS/TA | DATE

028

EXP. NUMBER	EXPERIMENT/SUBJECT		DATE	
NAME		LAB PARTNER	LOCKER/DESK NO.	COURSE & SECTION NO.

028

SIGNATURE		DATE	WITNESS/TA		DATE

EXP. NUMBER	EXPERIMENT/SUBJECT		DATE	
NAME		LAB PARTNER	LOCKER/DESK NO.	COURSE & SECTION NO.

029

SIGNATURE		DATE	WITNESS/TA		DATE

029

EXP. NUMBER	EXPERIMENT/SUBJECT		DATE		029
NAME		LAB PARTNER	LOCKER/DESK NO.	COURSE & SECTION NO.	

SIGNATURE		DATE	WITNESS/TA		DATE

029

EXP. NUMBER	EXPERIMENT/SUBJECT		DATE		030
NAME		LAB PARTNER	LOCKER/DESK NO.	COURSE & SECTION NO.	

SIGNATURE		DATE	WITNESS/TA		DATE

EXP. NUMBER	EXPERIMENT/SUBJECT		DATE		031
NAME		LAB PARTNER	LOCKER/DESK NO.	COURSE & SECTION NO.	

SIGNATURE		DATE	WITNESS/TA	DATE
				031

EXP. NUMBER	EXPERIMENT/SUBJECT		DATE	031
NAME		LAB PARTNER	LOCKER/DESK NO.	COURSE & SECTION NO.

SIGNATURE	DATE	WITNESS/TA	DATE

EXP. NUMBER	EXPERIMENT/SUBJECT		DATE	
NAME		LAB PARTNER	LOCKER/DESK NO.	COURSE & SECTION NO.

032

EXP. NUMBER	EXPERIMENT/SUBJECT		DATE		032
NAME		LAB PARTNER	LOCKER/DESK NO.	COURSE & SECTION NO.	

SIGNATURE		DATE	WITNESS/TA		DATE

EXP. NUMBER	EXPERIMENT/SUBJECT		DATE	033
NAME		LAB PARTNER	LOCKER/DESK NO.	COURSE & SECTION NO.

SIGNATURE		DATE	WITNESS/TA		DATE

EXP. NUMBER	EXPERIMENT/SUBJECT		DATE	
NAME		LAB PARTNER	LOCKER/DESK NO.	COURSE & SECTION NO.

EXP. NUMBER	EXPERIMENT/SUBJECT		DATE	
NAME		LAB PARTNER	LOCKER/DESK NO.	COURSE & SECTION NO.

034

SIGNATURE		DATE	WITNESS/TA		DATE

034

035

EXP. NUMBER	EXPERIMENT/SUBJECT		DATE	
NAME		LAB PARTNER	LOCKER/DESK NO.	COURSE & SECTION NO.

SIGNATURE		DATE	WITNESS/TA	DATE

EXP. NUMBER	EXPERIMENT/SUBJECT		DATE		036
NAME		LAB PARTNER	LOCKER/DESK NO.	COURSE & SECTION NO.	

SIGNATURE		DATE	WITNESS/TA		DATE

EXP. NUMBER	EXPERIMENT/SUBJECT		DATE		036
NAME		LAB PARTNER	LOCKER/DESK NO.	COURSE & SECTION NO.	

SIGNATURE		DATE	WITNESS/TA		DATE

EXP. NUMBER	EXPERIMENT/SUBJECT		DATE		037
NAME		LAB PARTNER	LOCKER/DESK NO.	COURSE & SECTION NO.	

SIGNATURE	DATE	WITNESS/TA	DATE

EXP. NUMBER	EXPERIMENT/SUBJECT		DATE		038
NAME		LAB PARTNER	LOCKER/DESK NO.	COURSE & SECTION NO.	

SIGNATURE		DATE	WITNESS/TA		DATE

EXP. NUMBER	EXPERIMENT/SUBJECT		DATE		038
NAME		LAB PARTNER	LOCKER/DESK NO.	COURSE & SECTION NO.	

SIGNATURE		DATE	WITNESS/TA		DATE
					038

EXP. NUMBER	EXPERIMENT/SUBJECT		DATE		039
NAME		LAB PARTNER	LOCKER/DESK NO.	COURSE & SECTION NO.	

SIGNATURE	DATE	WITNESS/TA	DATE

EXP. NUMBER	EXPERIMENT/SUBJECT		DATE	
NAME		LAB PARTNER	LOCKER/DESK NO.	COURSE & SECTION NO.

SIGNATURE		DATE	WITNESS/TA		DATE

EXP. NUMBER	EXPERIMENT/SUBJECT		DATE	
NAME		LAB PARTNER	LOCKER/DESK NO.	COURSE & SECTION NO.

040

SIGNATURE		DATE	WITNESS/TA		DATE

040

EXP. NUMBER	EXPERIMENT/SUBJECT		DATE	040
NAME		LAB PARTNER	LOCKER/DESK NO.	COURSE & SECTION NO.

SIGNATURE		DATE	WITNESS/TA		DATE

EXP. NUMBER	EXPERIMENT/SUBJECT		DATE	
NAME		LAB PARTNER	LOCKER/DESK NO.	COURSE & SECTION NO.

SIGNATURE		DATE	WITNESS/TA		DATE

EXP. NUMBER	EXPERIMENT/SUBJECT		DATE		041
NAME		LAB PARTNER	LOCKER/DESK NO.	COURSE & SECTION NO.	

SIGNATURE	DATE	WITNESS/TA	DATE

EXP. NUMBER	EXPERIMENT/SUBJECT		DATE	
NAME		LAB PARTNER	LOCKER/DESK NO.	COURSE & SECTION NO.

SIGNATURE		DATE	WITNESS/TA	DATE

EXP. NUMBER	EXPERIMENT/SUBJECT		DATE		043
NAME		LAB PARTNER	LOCKER/DESK NO.	COURSE & SECTION NO.	

SIGNATURE		DATE	WITNESS/TA		DATE
					043

EXP. NUMBER	EXPERIMENT/SUBJECT		DATE		043
NAME		LAB PARTNER	LOCKER/DESK NO.	COURSE & SECTION NO.	

SIGNATURE		DATE	WITNESS/TA		DATE

EXP. NUMBER	EXPERIMENT/SUBJECT		DATE	
NAME		LAB PARTNER	LOCKER/DESK NO.	COURSE & SECTION NO.

044

SIGNATURE		DATE	WITNESS/TA	DATE

EXP. NUMBER	EXPERIMENT/SUBJECT		DATE	
NAME		LAB PARTNER	LOCKER/DESK NO.	COURSE & SECTION NO.

SIGNATURE	DATE	WITNESS/TA	DATE

EXP. NUMBER	EXPERIMENT/SUBJECT		DATE		046
NAME		LAB PARTNER	LOCKER/DESK NO.	COURSE & SECTION NO.	

SIGNATURE		DATE	WITNESS/TA		DATE

EXP. NUMBER	EXPERIMENT/SUBJECT		DATE	
NAME		LAB PARTNER	LOCKER/DESK NO.	COURSE & SECTION NO.

SIGNATURE		DATE	WITNESS/TA		DATE

EXP. NUMBER	EXPERIMENT/SUBJECT		DATE	
NAME		LAB PARTNER	LOCKER/DESK NO.	COURSE & SECTION NO.

SIGNATURE		DATE	WITNESS/TA	DATE

EXP. NUMBER	EXPERIMENT/SUBJECT		DATE	
NAME		LAB PARTNER	LOCKER/DESK NO.	COURSE & SECTION NO.

047

EXP. NUMBER	EXPERIMENT/SUBJECT		DATE	
NAME		LAB PARTNER	LOCKER/DESK NO.	COURSE & SECTION NO.

SIGNATURE		DATE	WITNESS/TA		DATE

EXP. NUMBER	EXPERIMENT/SUBJECT		DATE	
NAME		LAB PARTNER	LOCKER/DESK NO.	COURSE & SECTION NO.

SIGNATURE		DATE	WITNESS/TA		DATE

EXP. NUMBER	EXPERIMENT/SUBJECT		DATE	
NAME		LAB PARTNER	LOCKER/DESK NO.	COURSE & SECTION NO.

SIGNATURE		DATE	WITNESS/TA		DATE

EXP. NUMBER	EXPERIMENT/SUBJECT		DATE		050
NAME		LAB PARTNER	LOCKER/DESK NO.	COURSE & SECTION NO.	

SIGNATURE		DATE	WITNESS/TA		DATE
					050

EXP. NUMBER	EXPERIMENT/SUBJECT		DATE	
NAME		LAB PARTNER	LOCKER/DESK NO.	COURSE & SECTION NO.

SIGNATURE		DATE	WITNESS/TA		DATE

EXP. NUMBER	EXPERIMENT/SUBJECT		DATE	
NAME		LAB PARTNER	LOCKER/DESK NO.	COURSE & SECTION NO.

052

SIGNATURE	DATE	WITNESS/TA	DATE

EXP. NUMBER	EXPERIMENT/SUBJECT		DATE		053
NAME		LAB PARTNER	LOCKER/DESK NO.	COURSE & SECTION NO.	

SIGNATURE		DATE	WITNESS/TA		DATE

EXP. NUMBER	EXPERIMENT/SUBJECT		DATE	053
NAME		LAB PARTNER	LOCKER/DESK NO.	COURSE & SECTION NO.

SIGNATURE		DATE	WITNESS/TA	DATE

EXP. NUMBER	EXPERIMENT/SUBJECT		DATE		054
NAME		LAB PARTNER	LOCKER/DESK NO.	COURSE & SECTION NO.	

SIGNATURE	DATE	WITNESS/TA	DATE

EXP. NUMBER	EXPERIMENT/SUBJECT		DATE	
NAME		LAB PARTNER	LOCKER/DESK NO.	COURSE & SECTION NO.

055

EXP. NUMBER	EXPERIMENT/SUBJECT		DATE	055
NAME		LAB PARTNER	LOCKER/DESK NO.	COURSE & SECTION NO.

SIGNATURE		DATE	WITNESS/TA	DATE

COPY

EXP. NUMBER	EXPERIMENT/SUBJECT		DATE		056
NAME		LAB PARTNER	LOCKER/DESK NO.	COURSE & SECTION NO.	

SIGNATURE	DATE	WITNESS/TA	DATE
			056

EXP. NUMBER	EXPERIMENT/SUBJECT		DATE		057
NAME		LAB PARTNER	LOCKER/DESK NO.	COURSE & SECTION NO.	

SIGNATURE		DATE	WITNESS/TA		DATE
					057

EXP. NUMBER	EXPERIMENT/SUBJECT		DATE		057
NAME		LAB PARTNER	LOCKER/DESK NO.	COURSE & SECTION NO.	

SIGNATURE		DATE	WITNESS/TA		DATE

EXP. NUMBER	EXPERIMENT/SUBJECT		DATE	
NAME		LAB PARTNER	LOCKER/DESK NO.	COURSE & SECTION NO.

SIGNATURE		DATE	WITNESS/TA		DATE

EXP. NUMBER	EXPERIMENT/SUBJECT		DATE	
NAME		LAB PARTNER	LOCKER/DESK NO.	COURSE & SECTION NO.

059

SIGNATURE		DATE	WITNESS/TA		DATE

059

EXP. NUMBER	EXPERIMENT/SUBJECT		DATE	
NAME		LAB PARTNER	LOCKER/DESK NO.	COURSE & SECTION NO.

SIGNATURE	DATE	WITNESS/TA	DATE

EXP. NUMBER	EXPERIMENT/SUBJECT		DATE		
NAME		LAB PARTNER	LOCKER/DESK NO.	COURSE & SECTION NO.	

060

COPY

SIGNATURE		DATE	WITNESS/TA		DATE

EXP. NUMBER	EXPERIMENT/SUBJECT		DATE	
NAME		LAB PARTNER	LOCKER/DESK NO.	COURSE & SECTION NO.

SIGNATURE	DATE	WITNESS/TA	DATE

EXP. NUMBER	EXPERIMENT/SUBJECT		DATE	
NAME		LAB PARTNER	LOCKER/DESK NO.	COURSE & SECTION NO.

062

SIGNATURE		DATE	WITNESS/TA		DATE

EXP. NUMBER	EXPERIMENT/SUBJECT		DATE		062
NAME		LAB PARTNER	LOCKER/DESK NO.	COURSE & SECTION NO.	

SIGNATURE		DATE	WITNESS/TA		DATE

EXP. NUMBER	EXPERIMENT/SUBJECT		DATE	
NAME		LAB PARTNER	LOCKER/DESK NO.	COURSE & SECTION NO.

SIGNATURE		DATE	WITNESS/TA	DATE

EXP. NUMBER	EXPERIMENT/SUBJECT		DATE	
NAME		LAB PARTNER	LOCKER/DESK NO.	COURSE & SECTION NO.

063

SIGNATURE		DATE	WITNESS/TA		DATE

EXP. NUMBER	EXPERIMENT/SUBJECT		DATE	
NAME		LAB PARTNER	LOCKER/DESK NO.	COURSE & SECTION NO.

SIGNATURE		DATE	WITNESS/TA		DATE

EXP. NUMBER	EXPERIMENT/SUBJECT		DATE		064
NAME		LAB PARTNER	LOCKER/DESK NO.	COURSE & SECTION NO.	

SIGNATURE		DATE	WITNESS/TA		DATE

EXP. NUMBER	EXPERIMENT/SUBJECT		DATE	
	NAME	LAB PARTNER	LOCKER/DESK NO.	COURSE & SECTION NO.

065

EXP. NUMBER	EXPERIMENT/SUBJECT		DATE		065
NAME		LAB PARTNER	LOCKER/DESK NO.	COURSE & SECTION NO.	

SIGNATURE		DATE	WITNESS/TA		DATE

EXP. NUMBER	EXPERIMENT/SUBJECT		DATE	
NAME		LAB PARTNER	LOCKER/DESK NO.	COURSE & SECTION NO.

SIGNATURE		DATE	WITNESS/TA		DATE

EXP. NUMBER	EXPERIMENT/SUBJECT		DATE	
NAME		LAB PARTNER	LOCKER/DESK NO.	COURSE & SECTION NO.

067

SIGNATURE	DATE	WITNESS/TA	DATE

EXP. NUMBER	EXPERIMENT/SUBJECT		DATE	
NAME		LAB PARTNER	LOCKER/DESK NO.	COURSE & SECTION NO.

SIGNATURE		DATE	WITNESS/TA		DATE

EXP. NUMBER	EXPERIMENT/SUBJECT		DATE		068
NAME		LAB PARTNER	LOCKER/DESK NO.	COURSE & SECTION NO.	

SIGNATURE		DATE	WITNESS/TA		DATE

EXP. NUMBER	EXPERIMENT/SUBJECT		DATE		069
NAME		LAB PARTNER	LOCKER/DESK NO.	COURSE & SECTION NO.	

SIGNATURE		DATE	WITNESS/TA		DATE

EXP. NUMBER	EXPERIMENT/SUBJECT		DATE	
NAME		LAB PARTNER	LOCKER/DESK NO.	COURSE & SECTION NO.

SIGNATURE	DATE	WITNESS/TA	DATE

EXP. NUMBER	EXPERIMENT/SUBJECT		DATE	
NAME		LAB PARTNER	LOCKER/DESK NO.	COURSE & SECTION NO.

071

SIGNATURE	DATE	WITNESS/TA	DATE

071

EXP. NUMBER	EXPERIMENT/SUBJECT		DATE	
NAME		LAB PARTNER	LOCKER/DESK NO.	COURSE & SECTION NO.

SIGNATURE	DATE	WITNESS/TA	DATE

EXP. NUMBER	EXPERIMENT/SUBJECT		DATE	
NAME		LAB PARTNER	LOCKER/DESK NO.	COURSE & SECTION NO.

SIGNATURE		DATE	WITNESS/TA	DATE

EXP. NUMBER	EXPERIMENT/SUBJECT		DATE		073
NAME		LAB PARTNER	LOCKER/DESK NO.	COURSE & SECTION NO.	

SIGNATURE		DATE	WITNESS/TA		DATE

EXP. NUMBER	EXPERIMENT/SUBJECT		DATE	
NAME		LAB PARTNER	LOCKER/DESK NO.	COURSE & SECTION NO.

SIGNATURE		DATE	WITNESS/TA	DATE

EXP. NUMBER	EXPERIMENT/SUBJECT		DATE		074
NAME		LAB PARTNER	LOCKER/DESK NO.	COURSE & SECTION NO.	

SIGNATURE		DATE	WITNESS/TA		DATE
					074

EXP. NUMBER	EXPERIMENT/SUBJECT		DATE	
NAME		LAB PARTNER	LOCKER/DESK NO.	COURSE & SECTION NO.

SIGNATURE		DATE	WITNESS/TA		DATE

EXP. NUMBER	EXPERIMENT/SUBJECT		DATE	
NAME		LAB PARTNER	LOCKER/DESK NO.	COURSE & SECTION NO.

076

SIGNATURE	DATE	WITNESS/TA	DATE

COPY

077

EXP. NUMBER	EXPERIMENT/SUBJECT		DATE	
NAME		LAB PARTNER	LOCKER/DESK NO.	COURSE & SECTION NO.

SIGNATURE		DATE	WITNESS/TA		DATE

EXP. NUMBER	EXPERIMENT/SUBJECT		DATE	
NAME		LAB PARTNER	LOCKER/DESK NO.	COURSE & SECTION NO.

SIGNATURE	DATE	WITNESS/TA	DATE

EXP. NUMBER	EXPERIMENT/SUBJECT		DATE		079
NAME		LAB PARTNER	LOCKER/DESK NO.	COURSE & SECTION NO.	

SIGNATURE		DATE	WITNESS/TA		DATE

EXP. NUMBER	EXPERIMENT/SUBJECT		DATE		080
NAME		LAB PARTNER	LOCKER/DESK NO.	COURSE & SECTION NO.	

SIGNATURE		DATE	WITNESS/TA		DATE

EXP. NUMBER	EXPERIMENT/SUBJECT		DATE	081
NAME		LAB PARTNER	LOCKER/DESK NO.	COURSE & SECTION NO.

SIGNATURE	DATE	WITNESS/TA	DATE

EXP. NUMBER	EXPERIMENT/SUBJECT		DATE	
NAME		LAB PARTNER	LOCKER/DESK NO.	COURSE & SECTION NO.

SIGNATURE		DATE	WITNESS/TA		DATE

EXP. NUMBER	EXPERIMENT/SUBJECT		DATE	
NAME		LAB PARTNER	LOCKER/DESK NO.	COURSE & SECTION NO.

082

SIGNATURE		DATE	WITNESS/TA		DATE

EXP. NUMBER	EXPERIMENT/SUBJECT		DATE	
NAME		LAB PARTNER	LOCKER/DESK NO.	COURSE & SECTION NO.

SIGNATURE		DATE	WITNESS/TA		DATE

EXP. NUMBER	EXPERIMENT/SUBJECT		DATE		083
NAME		LAB PARTNER	LOCKER/DESK NO.	COURSE & SECTION NO.	

SIGNATURE		DATE	WITNESS/TA		DATE

EXP. NUMBER	EXPERIMENT/SUBJECT		DATE	
NAME		LAB PARTNER	LOCKER/DESK NO.	COURSE & SECTION NO.

SIGNATURE	DATE	WITNESS/TA	DATE

EXP. NUMBER	EXPERIMENT/SUBJECT		DATE	
				084
NAME		LAB PARTNER	LOCKER/DESK NO.	COURSE & SECTION NO.

SIGNATURE		DATE	WITNESS/TA		DATE

EXP. NUMBER	EXPERIMENT/SUBJECT		DATE		085
NAME		LAB PARTNER	LOCKER/DESK NO.	COURSE & SECTION NO.	

SIGNATURE		DATE	WITNESS/TA		DATE
					085

EXP. NUMBER	EXPERIMENT/SUBJECT		DATE	085
NAME		LAB PARTNER	LOCKER/DESK NO.	COURSE & SECTION NO.

SIGNATURE	DATE	WITNESS/TA	DATE

EXP. NUMBER	EXPERIMENT/SUBJECT		DATE	
NAME		LAB PARTNER	LOCKER/DESK NO.	COURSE & SECTION NO.

SIGNATURE		DATE	WITNESS/TA		DATE

EXP. NUMBER	EXPERIMENT/SUBJECT		DATE		086
NAME		LAB PARTNER	LOCKER/DESK NO.	COURSE & SECTION NO.	

SIGNATURE		DATE	WITNESS/TA		DATE

EXP. NUMBER	EXPERIMENT/SUBJECT		DATE	
NAME		LAB PARTNER	LOCKER/DESK NO.	COURSE & SECTION NO.

SIGNATURE		DATE	WITNESS/TA		DATE

EXP. NUMBER	EXPERIMENT/SUBJECT		DATE	
NAME		LAB PARTNER	LOCKER/DESK NO.	COURSE & SECTION NO.

SIGNATURE		DATE	WITNESS/TA		DATE

EXP. NUMBER	EXPERIMENT/SUBJECT		DATE	088
NAME		LAB PARTNER	LOCKER/DESK NO.	COURSE & SECTION NO.

SIGNATURE	DATE	WITNESS/TA	DATE
			088

EXP. NUMBER	EXPERIMENT/SUBJECT		DATE	
NAME		LAB PARTNER	LOCKER/DESK NO.	COURSE & SECTION NO.

SIGNATURE		DATE	WITNESS/TA		DATE

EXP. NUMBER	EXPERIMENT/SUBJECT		DATE		089
NAME		LAB PARTNER	LOCKER/DESK NO.	COURSE & SECTION NO.	

SIGNATURE	DATE	WITNESS/TA	DATE

EXP. NUMBER	EXPERIMENT/SUBJECT		DATE	
NAME		LAB PARTNER	LOCKER/DESK NO.	COURSE & SECTION NO.

SIGNATURE	DATE	WITNESS/TA	DATE

EXP. NUMBER	EXPERIMENT/SUBJECT		DATE		090
NAME		LAB PARTNER	LOCKER/DESK NO.	COURSE & SECTION NO.	

SIGNATURE		DATE	WITNESS/TA		DATE	

EXP. NUMBER	EXPERIMENT/SUBJECT		DATE	
NAME		LAB PARTNER	LOCKER/DESK NO.	COURSE & SECTION NO.

SIGNATURE	DATE	WITNESS/TA	DATE

EXP. NUMBER	EXPERIMENT/SUBJECT		DATE		091
NAME		LAB PARTNER	LOCKER/DESK NO.	COURSE & SECTION NO.	

SIGNATURE		DATE	WITNESS/TA		DATE

EXP. NUMBER	EXPERIMENT/SUBJECT		DATE	
NAME		LAB PARTNER	LOCKER/DESK NO.	COURSE & SECTION NO.

SIGNATURE		DATE	WITNESS/TA		DATE

EXP. NUMBER	EXPERIMENT/SUBJECT	DATE	092
NAME	LAB PARTNER	LOCKER/DESK NO.	COURSE & SECTION NO.

SIGNATURE	DATE	WITNESS/TA	DATE

EXP. NUMBER	EXPERIMENT/SUBJECT		DATE	
NAME		LAB PARTNER	LOCKER/DESK NO.	COURSE & SECTION NO.

SIGNATURE	DATE	WITNESS/TA	DATE

EXP. NUMBER	EXPERIMENT/SUBJECT		DATE		093
NAME		LAB PARTNER	LOCKER/DESK NO.	COURSE & SECTION NO.	

COPY

SIGNATURE		DATE	WITNESS/TA		DATE

EXP. NUMBER	EXPERIMENT/SUBJECT		DATE		094
NAME		LAB PARTNER	LOCKER/DESK NO.	COURSE & SECTION NO.	

SIGNATURE		DATE	WITNESS/TA		DATE 094

EXP. NUMBER	EXPERIMENT/SUBJECT		DATE	
NAME		LAB PARTNER	LOCKER/DESK NO.	COURSE & SECTION NO.

094

SIGNATURE		DATE	WITNESS/TA		DATE

EXP. NUMBER	EXPERIMENT/SUBJECT		DATE	
NAME		LAB PARTNER	LOCKER/DESK NO.	COURSE & SECTION NO.

095

SIGNATURE		DATE	WITNESS/TA	DATE

EXP. NUMBER	EXPERIMENT/SUBJECT		DATE	
NAME		LAB PARTNER	LOCKER/DESK NO.	COURSE & SECTION NO.

SIGNATURE		DATE	WITNESS/TA	DATE

EXP. NUMBER	EXPERIMENT/SUBJECT		DATE	
NAME		LAB PARTNER	LOCKER/DESK NO.	COURSE & SECTION NO.

SIGNATURE		DATE	WITNESS/TA		DATE

EXP. NUMBER	EXPERIMENT/SUBJECT		DATE	
NAME		LAB PARTNER	LOCKER/DESK NO.	COURSE & SECTION NO.

SIGNATURE		DATE	WITNESS/TA		DATE

EXP. NUMBER	EXPERIMENT/SUBJECT		DATE		097
NAME		LAB PARTNER	LOCKER/DESK NO.	COURSE & SECTION NO.	

SIGNATURE		DATE	WITNESS/TA		DATE

EXP. NUMBER	EXPERIMENT/SUBJECT		DATE	
NAME		LAB PARTNER	LOCKER/DESK NO.	COURSE & SECTION NO.

098

EXP. NUMBER	EXPERIMENT/SUBJECT		DATE		099
NAME		LAB PARTNER	LOCKER/DESK NO.	COURSE & SECTION NO.	

SIGNATURE		DATE	WITNESS/TA		DATE

EXP. NUMBER	EXPERIMENT/SUBJECT		DATE		099
NAME		LAB PARTNER	LOCKER/DESK NO.	COURSE & SECTION NO.	

SIGNATURE		DATE	WITNESS/TA		DATE

EXP. NUMBER	EXPERIMENT/SUBJECT		DATE		100
NAME		LAB PARTNER	LOCKER/DESK NO.	COURSE & SECTION NO.	

SIGNATURE		DATE	WITNESS/TA		DATE